D1747593

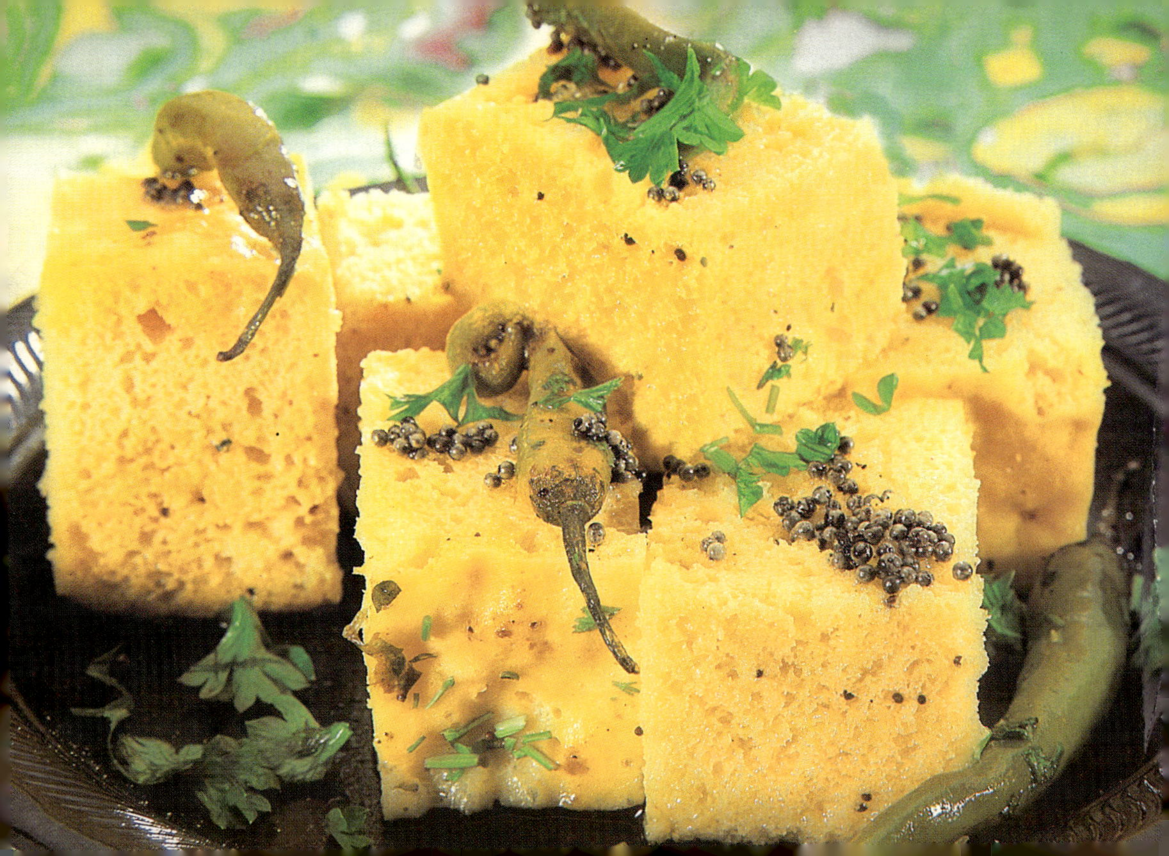

Nita Mehta's
NEW MICROWAVE
Cookbook

100% Vegetarian

Nita Mehta

B.Sc. (Home Science), M.Sc. (Food and Nutrition), Gold Medalist

TANYA MEHTA

SNAB
Excellence in Books

Nita Mehta's
NEW MICROWAVE
Cookbook

© Copyright 2004-2009 **SNAB** Publishers Pvt Ltd

WORLD RIGHTS RESERVED. The contents—all recipes, photographs and drawings are original and copyrighted. No portion of this book shall be reproduced, stored in a retrieval system or transmitted by any means, electronic, mechanical, photocopying, recording or otherwise, without the written permission of the publishers.

While every precaution is taken in the preparation of this book, the publisher and the author assume no responsibility for errors or omissions. Neither is any liability assumed for damages resulting from the use of information contained herein.

TRADEMARKS ACKNOWLEDGED. Trademarks used, if any, are acknowledged as trademarks of their respective owners. These are used as reference only and no trademark infringement is intended upon.

4th Print 2009
ISBN 978-81-7869-076-6

Food Styling and Photography: **SNAB**

Layout and laser typesetting :

N.I.T.A. National Information Technology Academy
3A/3, Asaf Ali Road
New Delhi-110002
☎ 23252948

Published by :

SNAB
Excellence in Books
Publishers Pvt. Ltd.
3A/3 Asaf Ali Road,
New Delhi - 110002
Tel: 23252948, 23250091
Telefax:91-11-23250091

Editorial and Marketing office:
E-159, Greater Kailash-II, N.Delhi-48
Fax: 91-11-29225218, 29229558
Tel: 91-11-29214011, 29218727, 29218574
E-Mail: nitamehta@email.com
nitamehta@nitamehta.com
Website: http://www.nitamehta.com
Website: http://www.snabindia.com

Contributing Writers :
Anurag Mehta
Subhash Mehta

Editorial & Proofreading :
Rakesh
Ramesh

Distributed by :

THE VARIETY BOOK DEPOT
A.V.G. Bhavan, M 3 Con Circus,
New Delhi - 110 001
Tel : 23417175, 23412567; Fax : 23415335
Email: varietybookdepot@rediffmail.com

Printed by :
DEVTECH PUBLISHERS & PRINTERS PVT LTD

Rs. 89/-

Picture on cover:	*Paneer Makhani*	*Recipe on page 58*
	Ghiya Channe ki Dal	*Recipe on page 42*
Picture on page 1:	*Crispy Achaari Mirch*	*Recipe on page 70*
	Grilled Besani Subzi	*Recipe on page 63*
Picture on page 2:	*Instant Khaman Dhokla*	*Recipe on page 32*
Picture on back cover:	*Eggless Cake with Mocha Icing* ...	*Recipe on page 99*

Introduction

The microwave helps today's women facing time constraints, to prepare a variety of favourite delicacies in a faster and a simpler manner. It leaves her with more time to spend with the family. Microwave makes the cooking simpler as the food does not stick or burn and hence it does not need constant stirring.

This book covers a range of vegetarian recipes, starting from starters to soups to main course Indian, Continental, Chinese and Thai dishes. The recipes have been adapted to suit the Indian palate.

Look forward to these wonderful recipes and share it with those you love and care about!

Nita Mehta

About The Recipes

what's in a cup?

INDIAN CUP
1 teacup = 200 ml liquid
AMERICAN CUP
1 cup = 240 ml liquid (8 oz.)

The recipes in this book were tested with the Indian teacup which holds 200 ml liquid.

Contents

Introduction   6
Basics of Microwave Cooking   10
Microwave Tips   12
Interesting Uses of a Microwave   14
Utensils used in the Microwave Oven   17

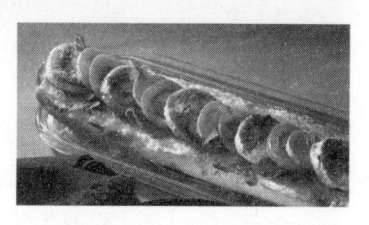

SNACKS   22

Soya Kebabs   23
Bean Squares   24
Pav Bhaji   26

Paneer Tikka   28
Instant Khaman Dhokla   32
Tomato-Kaju Idli   34

SOUP 36

Sweet Corn Soup 36   Capsicum Soup 38

INDIAN CURRIES 39

Palak Paneer 40	Mixed Veggie Curry 52
Ghiya-Channe ki Dal 42	Khumb Matar Miloni 54
Carrot Kofta Curry 44	Special Sambar 56
Makai-Mirch Salan 46	Paneer Makhani 58
Paneer Pista Haryali 50	Chawal Bhare Tamatar 60

INDIAN - DRY & MASALA 62

Dal Maharani 62	Crispy Achaari Mirch 70
Grilled Besani Subzi 63	Zayekedar Arbi 72
Mili-Juli-Vegetable 64	Baingan ka Bharta 74
Achaari Bhindi 68	Anjeeri Gobhi 77
	Kadhai Paneer 78

RICE 79

Steamed Rice 79 Subz Pullao 80

CHINESE AND THAI 82

Honey Chilli Veggies 82 Veggie Thai Red Curry 88
Broccoli in Butter Sauce 84 Glass Noodles with
Sesame Paste 90

BAKED DISHES 92

Macaroni Alfredo 92 Vegetable au Gratin 94
Rice-Vegetable Ring 96

DESSERTS AND CAKES 98

Gajar ka Halwa 98 Chocolate Walnut Cake
Eggless Cake with (Eggless) 101
Mocha Icing 99 Vanilla Cake 102

Basics of Microwave Cooking

Timing: Set the timings carefully, foods can become hard and leathery, if overcooked. It is always better to undercook than to overcook in a microwave. The larger the volume of food there is, the more timing is needed to cook it. 4 potatoes cook in 6 minutes, whereas 8 potatoes cook in about 9 minutes. Therefore, if the quantity in a recipe is changed, an adjustment in timing is necessary. When doubling a recipe, increase the cooking time 50% approximately and when cutting a recipe in half, reduce time by about 40%.

Standing time: Food continues to cook for sometime, even after it is removed from the microwave. For example, the cake cooked in a microwave looks very moist and undone when removed from the oven after microwaving it for the specified time, but after it is left aside for 8-10 minutes, it turns perfect.

Covering: Covers are used to trap steam, prevent dehydration, speed cooking time, and help food retain its natural moisture. When covering with paper napkins, a good microwave cooking practice, be sure to use a double width that will enable you to tuck the paper under the bottom of the cooking dish. Otherwise, it will tend to rise off the dish due to the air movement. A handy idea to keep in mind; a heatproof china plate is a good substitute for a lid. For shorter cooking time (within 6 minutes) cling films can also be used.

Stirring: If necessary, stir from the outside to the center because the outside area heats faster than the center when microwaves are in use. Stirring blends the flavours and promotes even heating. Stir only as directed in the recipes, constant stirring is never required, frequent stirring is rare.

Arrangement: The microwaves always penetrate the outer portion of food first, so food should be arranged with the thicker areas near the edge of the dish and the thinner portions near the center. Chicken/Mutton should be so placed that the meaty part is towards the outside. Food such as tomatoes, potatoes and corn should be arranged in a circle, rather than in rows.

Microwave Tips

- Never over-cook food as it becomes tough and leathery. Give the dish a little standing time before you test it, to avoid over cooking.

- Never pile food on top of each other. It cooks better, evenly and quickly when spaced apart.

- Food cooks better in a round container than in a square one. In square or rectangular bowls, the food gets overcooked at the corners.

- Do not add salt at the time of starting the cooking as it leads to increase in the cooking time.

- Do not add more water than required, however a little water must be added to prevent dehydration of the vegetables. When the vegetables get dehydrated, there is a loss of natural juices as well. But addition of extra water increases the cooking time.

- Do not cook eggs in their shells (pressure will cause them to explode).

- Do not deep fry in a microwave (the temperature of oil cannot be controlled).

- Do not cook and reheat puddings having alcohol (they can easily catch fire).

- Do not use containers with restricted openings, such as bottles.

- Use deep dishes to prepare gravies, filling the dish only ¾ to avoid spillage.

- Do not use aluminium foil for covering dishes in the microwave mode. Do not reheat foods (sweets like *ladoos, burfi* etc.) with silver sheet, as it leads to sparking.

- When using the convec mode put the dish on the wire rack to get even baking.

- Always preheat the oven when you want to use the convec mode. Grilling does not need preheating.

- When making tikkas or other tandoori delicacies cover the plate beneath the rack with aluminium foil to collect the drippings.

Interesting Uses of a Microwave

- **Making ghee**. Keep 1½ - 2 cups malai (milk topping) in a big glass bowl and microwave on high for 15-20 minutes to get desi ghee without burning your kadhai (wok). Stir once or twice inbetween.

- **Blanching almonds** to remove skin. Put almonds in a small bowl of water and microwave for 3 minutes or till water just starts to boil. After the water cools, the almonds can be peeled very easily.

- **Freshening** stale chips, biscuits or cornflakes. Place the chips or biscuits in a napkin, uncovered, for about 1 minute per bowl or until they feel warm. Wait for a few minutes to allow cooling and serve.

- **Boiling** (actually microwaving) potatoes. Wash potatoes and put them in a polythene bag. Microwave high for 5 minutes for 4 medium potatoes.

- **Making** khatti mithi chutney. Mix 1 tbsp amchur, 3 tbsp sugar, some water along with spices in a glass bowl. Microwave, stirring in between.

- **Warming** baby's milk bottle. Do check the temperature of the milk on your inner wrist. The bottle will not become hot, while the milk will.

- **Softening** too-hard ice cream, cream, cheese and butter.

- **Making** dry bread crumbs from fresh bread. Crumble the slice of bread and microwave the bits of slices for 2-3 minutes. Mix once and microwave further for another minute or two. Give some standing time to the moist bread to dry out and then grind in a mixer to get crumbs.

- **Drying** herbs. Fresh parsley, dill (soye), mint (poodina), coriander (dhaniyan), fenugreek greens (methi) — all greens can be dried in a microwave, preserving the green colour. Give them some standing time to turn dry. Use them in raitas and curries.

- **Melting** chocolate, butter, jam, honey, etc. Dissolving gelatine.

- **Sterilizing** jars for storing home made jams and pickles.

- **Freshening** stale bread by placing 2 slices between the folds of a paper and microwaving for 20 seconds. It turns absolutely soft and the stale bread becomes perfect for sandwiches.

- To **roast** 1 tbsp of cashews spread on a microproof plate and microwave for 1 minute to get golden roasted cashews.
- To **roast** papad place 2 papads on a paper napkin and microwave for 1½ minutes. Turn side once inbetween.
- To **blanch** 4 tomatoes, put a cross at the stem end of each tomato. Place tomatoes on a microproof plate and microwave for 2 minutes. Peel after they cold down.
- To **cook** corn, wash a corn on the cob and place in a plastic bag microwaver for 2-3 minutes to get soft corn.
- To **boil** ½ kg arbi wash and put in a plastic bag. Microwave for 11 minutes, turning once inbetween.

Utensils used in the Microwave Oven

MODE	CAN USE	DO NOT USE
Microwave Round or oval dishes are better for this mode as the corners of the square dish absorb more microwave energy or rays and hence food at the corners tends to get over cooked.	China Pottery (earthenware) Heatproof glass dishes like pyrex, borosil etc. Paper and cloth napkins as covers Plastic or cling wrap can be used as cover for short durations. Wooden skewers and toothpicks Plastic or polythene cooking bags.	China or any other utensil with gold or silver lining. Very delicate glass dishes Metal cake tins or any other metal Aluminium foil as covers Metal skewers

MODE	CAN USE	DO NOT USE
Convection In this mode there are no microwave rays or microwave energy. The oven becomes a conventional oven when put on this mode so all utnesils which go in the regular oven work well in the microwave oven when set on the convection mode.	Metal cake tins or any other metal utensil Heat proof glass dishes like pyrex of borosil Metal skewers Aluminium foil as covers	Delicate glass dishes which are not heatproof Wooden skewers Paper and cloth napkins or plastic wraps
Grill In this mode there are no microwaves so all heatproof utensils work well.	All as given for convec mode	All as given for convec mode *Contd...*

Rice Vegetable Ring: Recipe on Page 96 ➤

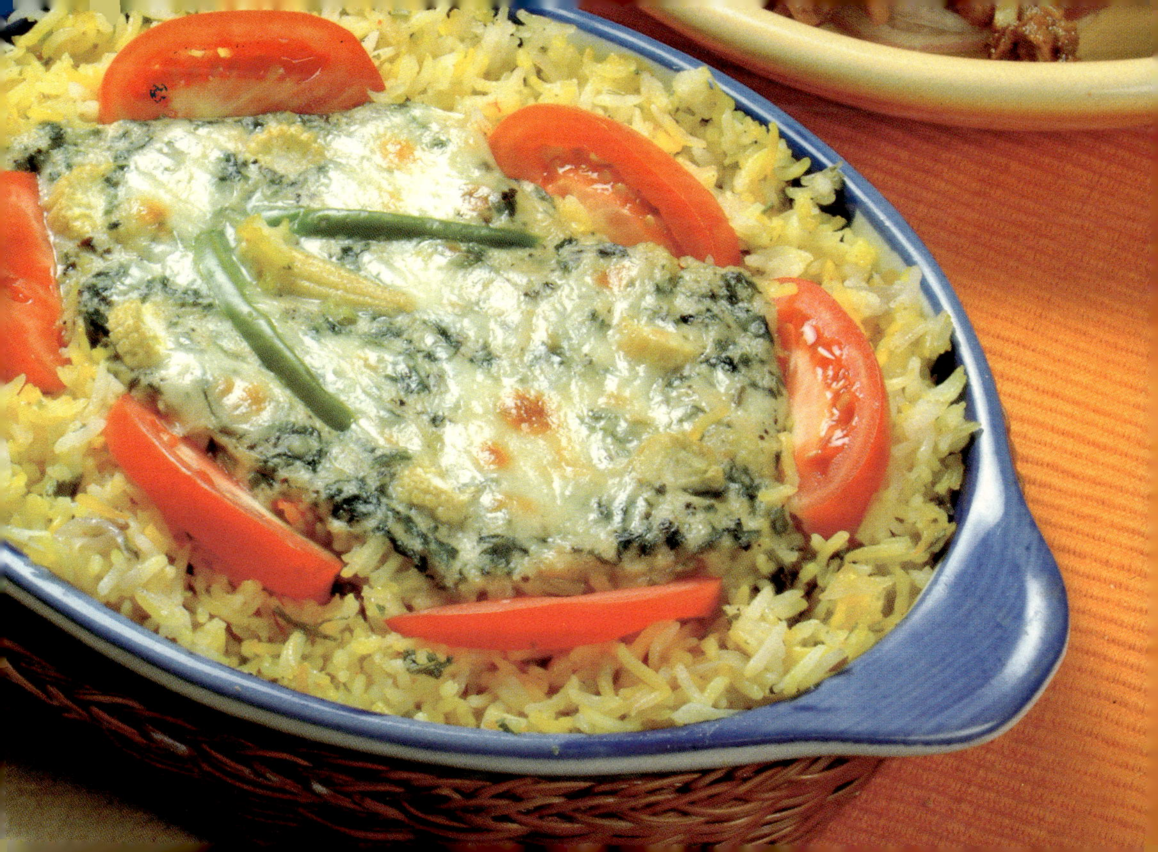

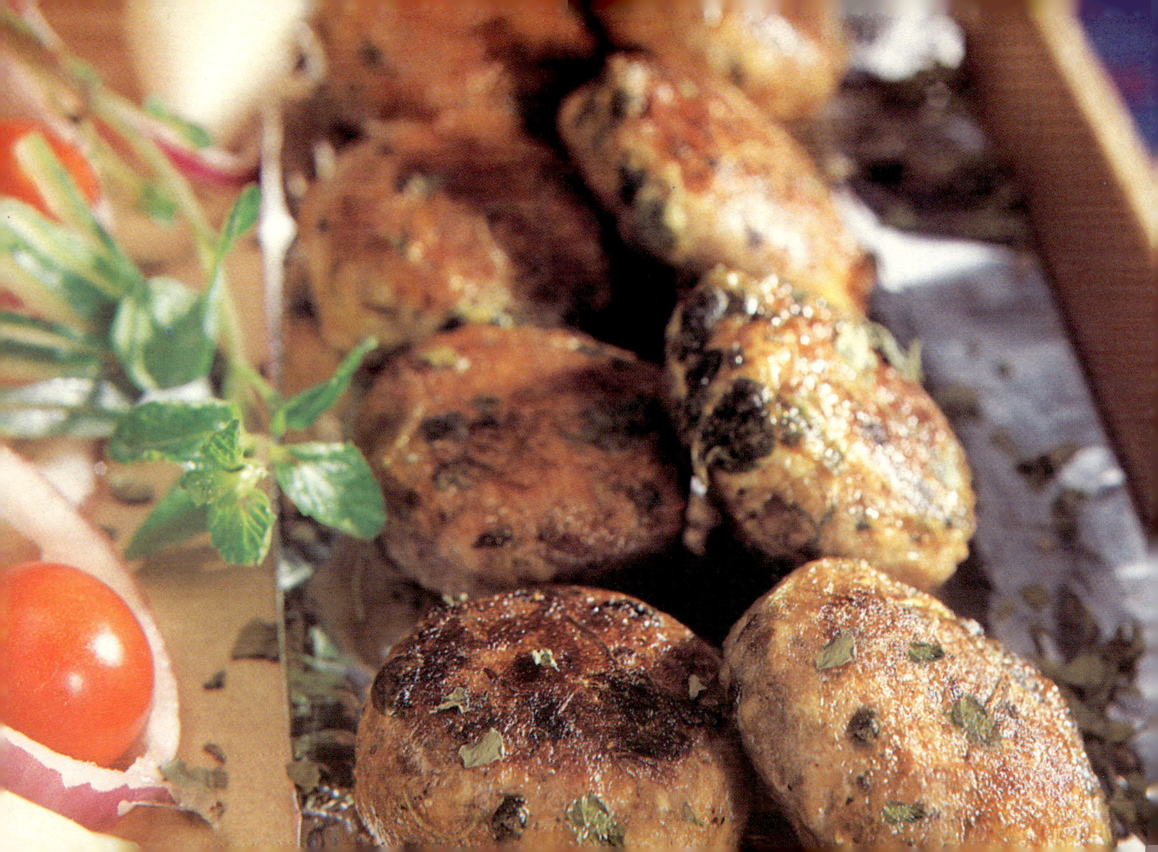

MODE	CAN USE	DO NOT USE
Combination (Micro+Oven) (Micro+Grill) Utensils must be microproof as well as heatproof for both the combination modes	Heat proof glass dishes like pyrex or borosil. Use a glass microproof and heatproof glass plate as cover	Metal tins, China Utensils Wooden & metal skewers Aluminium foil, paper or cloth napkins or plastic wrap.
Combination (Grill+Oven) Utensils must be heatproof	All as given for convec mode	All as for convec mode

◁ *Soya Kebabs : Recipe on Page 23*

Snacks

Soya Kebabs

Serves 4   *Picture on page 20*

1 cup soya granules (nutri nugget granules)
100 gm paneer - grated (1cup)
2 bread slices - torn in small pieces and churned in a mixer to get fresh crumbs
1 tsp garam masala, 1 tsp salt, ½ tsp red chilli powder
1½ tbsp tomato ketchup, 2 tbsp chopped green coriander - finely chopped
1 green chilli - chopped finely

1. Soak granules in 1 cup water for 15 minutes.
2. Strain, squeeze out the water well from the soya granules.
3. Add paneer, fresh bread crumbs, garam masala, salt, red chilli powder, tomato ketchup, chopped green coriander or mint and green chilli to the soya granules. Mix very well.
4. Make balls from it. Flatten each ball to get a kebab, with oiled hands.
5. Place on the grill of the oven. Spoon 1 tsp oil on each, grill for 20 minutes. After 10 minutes turn side in between and spoon 1 tsp oil on the other side. Continue grilling for the remaining 10 minutes. Serve

Bean Squares

A quick Mexican starter - crackers topped with cheesy beans and roasted peanuts.

Serves 4

8 cream cracker biscuits
½ cup grated cheese
½ cup boiled rajmah (red kidney beans)
2 tbsp tomato sauce
½ tsp salt, ¼ tsp red chilli powder
2 green chillies - deseeded, finely chopped
½ tsp oregano
a few roasted peanuts

SOUR CREAM
2 tbsp fresh cream - chilled
½ cup thick dahi (yogurt) - hang for 15 minutes in a muslin cloth & squeeze lightly
½ tsp lemon juice, ¼ tsp salt, or to taste
¼ tsp pepper, preferably white pepper

1. For the sour cream, hang curd in a muslin cloth for 15 minutes.
2. Beat the hung curd till smooth. Gently mix lemon juice, cream, salt and pepper. Keep sour cream in the refrigerator till serving time.
3. Mix cheese, boiled rajmah, tomato sauce, salt, red chilli powder, oregano & chopped green chillles. Keep topping aside till serving time.
4. To serve, spread 1 tbsp full of the bean topping on each biscuit in a heap, leaving the edges clean.
5. Place a paper napkin on the glass plate in the microwave.
6. Keep all the cream cracker biscuits together on it and microwave at 60% power for 3 minutes.
7. Serve each biscuit with a blob of sour cream and then top with a peanut.

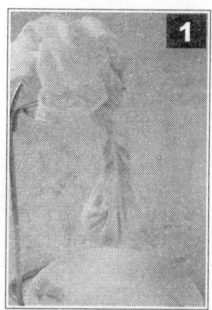

Pav Bhaji

Mixed vegetables flavoured with a fragrant spice blend. Enjoy it as a snack or for dinner.

Serves 4

3 onions - chopped finely
3 potatoes
2 carrots - peeled and chopped
½ cup peas
1½ cups chopped cauliflower
1 cup chopped cabbage
3 tbsp oil
3 tbsp butter
2 tsp ginger-garlic paste
2½ tbsp pav bhaji masala
¼ tsp haldi (turmeric powder)
1½ tsp salt
3 tomatoes - chopped
1 tbsp chopped coriander

1. Wash potatoes. Put in a plastic bag and microwave for 5 minutes. Peel and mash coarsely.
2. In a deep microproof bowl, put carrots, peas, cauliflower and cabbage. Add ½ cup water. Mix and microwave for 8 minutes. Let it cool. Blend roughly in a mixer for 1-2 seconds. Do not make it into a paste.
3. In a microproof dish add 3 tbsp oil, onions, ginger-garlic paste, 2 tbsp pav bhaji masala and haldi. Mix well. Microwave for 6 minutes.
4. Add tomatoes and the roughly mashed vegetables. Mix well. Add 2 tbsp butter, 1½ tsp salt. Cover and microwave for 10 minutes. Stir once in between.
5. Add 1 cup water. Mix and microwave for 5 minutes.
6. Add 1 tsp pav bhaji masala, 2 tbsp chopped coriander and 1 tbsp butter. Mix and serve.

Paneer Tikka

The universal Indian delight, now made more delicious!

Picture on facing page Serves 4

300 gm paneer - cut into 2" squares of about ¾" thickness
1 large capsicum - cut into 1" pieces or rings
1 onion - cut into 4 pieces
1 tomato - cut into 8 pieces

MARINADE

1 cup dahi - hang in a muslin cloth for 20 minutes
3 tbsp thick malai or thick cream
a few drops of orange colour or a pinch of haldi (turmeric)
1½ tbsp oil, 1 tbsp cornflour
½ tsp amchoor ½ tsp black salt
½ tsp red chilli powder, ¾ tsp salt
1 tbsp tandoori masala or ½ tsp garam masala
1 tbsp ginger-garlic paste

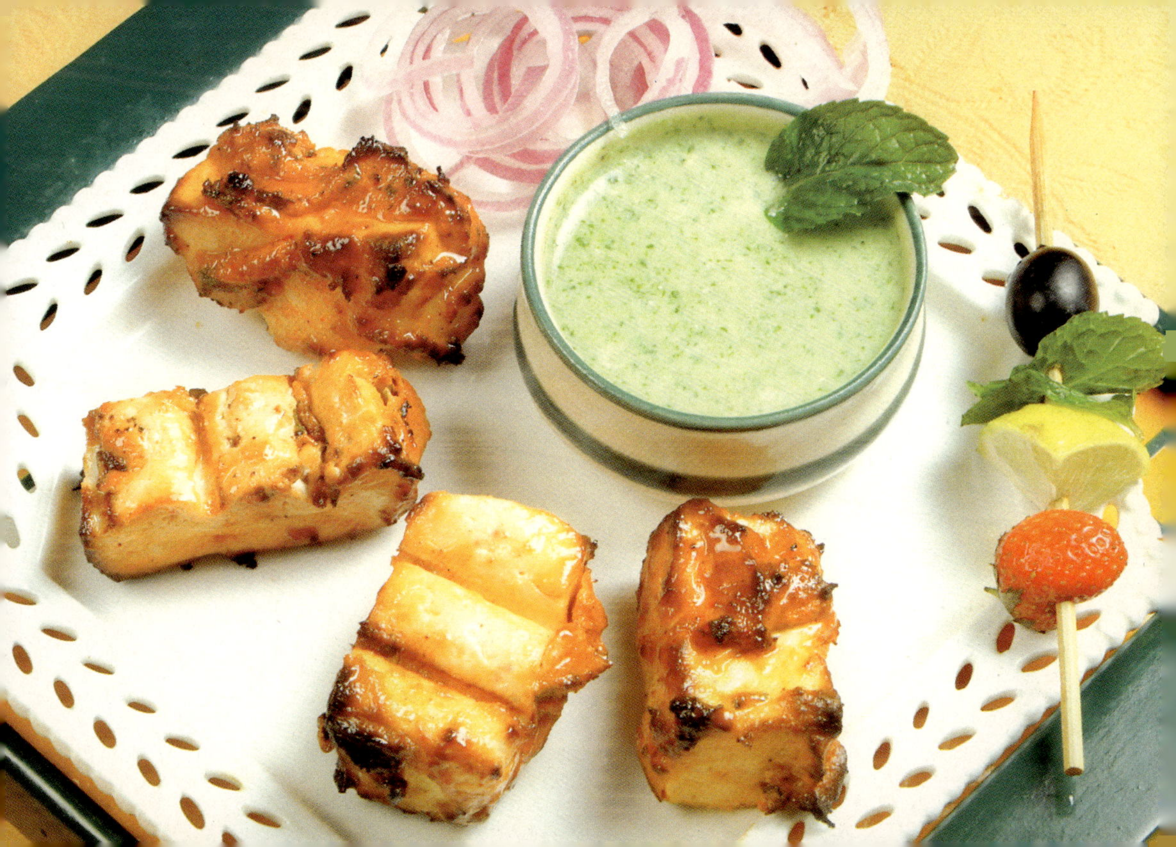

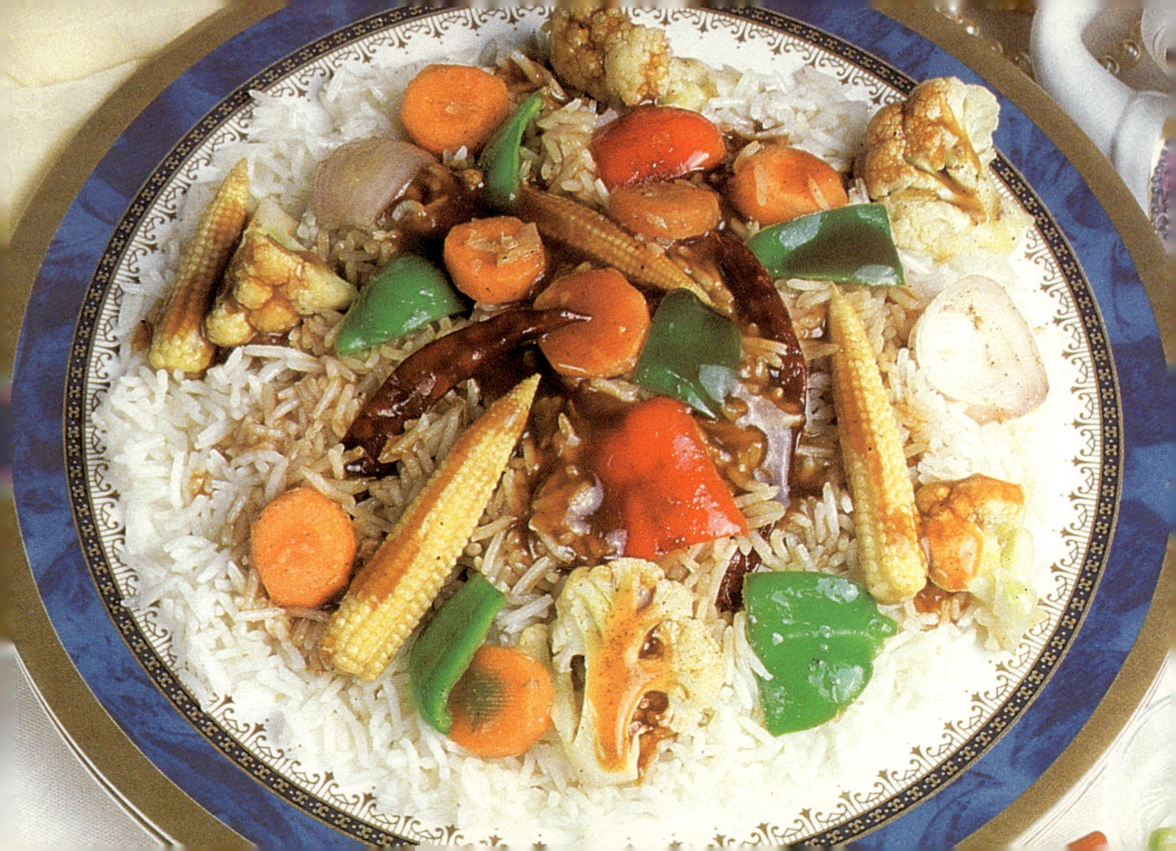

1. Mix all ingredients of the marinade in a bowl. Add paneer. Mix well.
2. Grease wire or grill rack. Arrange paneer on the greased wire rack.
3. After all the paneer pieces are done, put capsicum, onions and tomato together in the left over marinade and mix well to coat the vegetables. Place vegetables also on rack.
4. Set your microwave oven at 200°C using the oven (convection) mode and press start to preheat.
5. Put the tikkas in the hot oven.
6. Re-set the preheated oven again at 160°C for 20 minutes. Cook the tikkas for 15 minutes.
7. Spoon some melted butter on the tikkas and cook further for 5 minutes.
8. Remove from oven. To serve, transfer them to a micro-proof serving platter. Sprinkle chat masala and lemon juice. Microwave for 1 minute.

◁ *Honey Chilli Veggies : Recipe on Page 82*

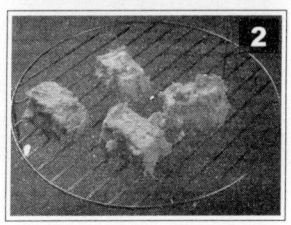

Instant Khaman Dhokla

This light Gujarati snack is quick to make in a microwave.

Picture on page 2   *Serves 6*

1½ cups besan (gram flour)
1 cup water, 1 tbsp oil
½ tsp haldi (turmeric)
1 tsp green chilli paste, 1 tsp ginger paste
1 tsp salt, 1 tsp sugar
¼ tsp soda-bi-carb (mitha soda)
1½ tsp eno fruit salt, 2 tsp lemon juice

TEMPERING/CHOWNK
2 tbsp oil, 1 tsp rai (mustard seeds)
2-3 green chillies - slit into long pieces
¼ cup white vinegar
¾ cup water, 1 tbsp sugar

1. Grease a 7" diameter round, flat dish with oil. Keep aside.
2. Sift besan through sieve to make it light and free of any lumps.
3. Mix besan, water, oil, turmeric powder, salt, sugar, chilli paste, ginger paste and water to a smooth batter.
4. Add eno fruit salt and soda-bi-carb to the batter and pour lemon juice over it. Beat well for a few seconds.
5. Immediately pour this mixture in the greased dish. Microwave uncovered for 6 minutes. Remove from oven and keep aside.
6. To temper, microwave oil, green chillies, rai, water, sugar and vinegar for 4½ minutes. Pour over the dhokla and wait for ½ hour to absorb it and to turn soft.
7. Cool and cut into 1½" pieces.
8. Sprinkle chopped coriander. Serve.

Tomato-Kaju Idli

Makes 6 idlis

1 cup suji (rawa)
1½ tbsp oil
1 cup curd
½ cup water, approx.
½ tsp soda-bicarb
¾ tsp salt

OTHER INGREDIENTS
1 firm tomato- cut into 8 slices
4-5 cashews - split into halves
8-10 curry leaves

1. In a dish put 1½ tbsp oil. Microwave for 1 minute.
2. Add suji. Mix well. Microwave uncovered for 2 minutes.
3. Add salt. Mix well. Allow to cool.
4. Add curd and water. Mix till smooth.
5. Add soda-bicarb. Mix very well till smooth. Keep aside for 10 minutes.
6. Grease 6 small glass katoris or plastic idli boxes. Arrange a slice of tomato, a split cashew half and a curry leaf at the bottom of the katori.
7. Pour 3-4 tbsp mixture into each katori.
8. Arrange katoris in a ring in the microwave and microwave uncovered for 3½ minutes. Do not microwave more even if they appear wet.
9. Let them stand for 5 minutes. They will turn dry. Serve hot with sambhar and chutney.

Soups

Sweet Corn Soup

The all time favourite Chinese soup!

Serves 6-7

½ tin cream style sweet corn tin (460 gm for full tin- 2½ cups)
¼ cup chopped cabbage
¼ cup grated carrot
1 spring onion - finely chopped, including the greens
1 tsp vinegar
1 tsp red chilli sauce
1 tbsp green chilli sauce
pinch of ajinomoto
¼ tsp pepper
¾ tsp salt, or to taste
2½ tbsp cornflour mixed in ¼ cup water

1. In a deep bowl, mix cream style corn (1¼ cups) with 4 cups water. Microwave for 8 minutes or more till it comes to a boil. Stir once inbetween.
2. Add all other ingredients and microwave again for 5 minutes. Stir once inbetween.
3. Microwave for 1-2 minutes more if the soup is not thick enough. Serve hot with green chillies in vinegar.

Note: The left over cream style corn can be stored in a box in the freezer compartment of the fridge for a month or even more.

Capsicum Soup

A cheesy light green soup.

Serves 4

4 medium sized capsicums - cut into big pieces
2 tomatoes - cut into big pieces
2 cups water, ½ cup milk, 2 tsp cheese spread
1 tsp salt , ½ tsp pepper or to taste, 1 tsp butter

1. Microwave capsicum and tomato with 1 cup water in a microproof bowl for 3 minutes.
2. Remove from the microwave, cool.
3. Add 1 cup water. Churn in a mixer to get a smooth puree. Strain puree.
4. To the strained puree add milk, cheese spread, salt, pepper and butter. Microwave for 6 minutes.
5. Pour into individual bowls and serve hot.

Indian Curries

Palak Paneer

Spinach and cottage cheese - a wonderful combination!

Serves 4

250 gm paneer - cut into 1" cubes
1 bundle (600 gm) spinach - break the leaves only, discarding the stem
2 tbsp oil
1 tsp cumin seeds (jeera)
½" piece ginger
4-5 flakes of garlic, 2 onions - chopped
1 green chilli - chopped
2 tsp coriander powder (dhania powder)
½ tsp garam masala
2 tomatoes - chopped
2 tbsp dry fenugreek leaves (kasoori methi)
1 tbsp butter
½ tsp red chilli powder
¾ tsp salt, ¼ tsp sugar
¼ cup milk

1. In a microproof deep bowl put oil, jeera, ginger, garlic, onions, green chilli, 2 tsp dhania powder and garam masala. Mix well. Microwave for 5 minutes.
2. Add chopped tomatoes and kasoori methi. Mix well. Add washed spinach leaves. Microwave for 8 minutes.
3. Cool spinach. Blend with ½ cup water.
4. Transfer the spinach puree to the same microproof dish. Add 1 tbsp butter, ½ tsp red chilli powder, paneer, ¾ tsp salt, ¼ tsp sugar, ¼ cup milk and ¼ cup water. Mix well.
5. Microwave for 5 minutes. Serve hot.

Ghiya-Channe ki Dal

Gram lentils cooked with bottle gourd and tempered to perfection.

Picture on cover *Serves 4*

¾ cup channe ki dal (gram lentils) - washed and soaked for ½ hour
½ small (200 gms) ghiya (bottle gourd) - peeled and chopped
1 tsp salt
½ tsp haldi (turmeric powder)
2 tsp desi ghee or oil
½ tsp red chilli powder

TOMATO-ONION BAGHAR

3 tbsp oil
1 tsp cumin seeds (jeera)
1 onion - finely sliced, 1 tomato - finely chopped
2 tbsp chopped coriander
2 green chillies
1 tsp dhania powder
½ tsp garam masala, ½ tsp amchoor
½ tsp red chilli powder

1. Pick, clean and wash dal. Soak for ½ hour.
2. Drain water from dal. Mix dal, ghiya, salt, haldi, desi ghee, red chilli powder and 2 cups water in a deep bowl. Microwave covered for 6 minutes.
3. Stir once inbetween. Remove cover and microwave for 20 minutes or till dal turns soft. Mash lightly. Cover and keep aside.
4. For the baghar, mix oil with jeera in a microproof dish. Microwave for 2 minutes. Add onions and microwave for 4 minutes till golden.
5. Add tomato, coriander, whole green chillies, dhania powder, garam masala, amchoor and red chilli powder. Mix well. Microwave for 3 minutes. Pour over the hot cooked dal. Mix gently. Serve hot.

Carrot Kofta Curry

Carrot balls stuffed with raisins in a simple, yet tasty curry.

Serves 4

3 tbsp oil
2 onions - ground to a paste in a mixer
2 tomatoes - pureed in a mixer
2 tsp dhania powder, ¼ tsp haldi
¼ tsp garam masala
¼ tsp red chilli powder

KOFTE

2 carrots - grated
2 bread slices - break into pieces and grind to crumbs in a mixer
1 green chilli - chopped
1 tsp ginger paste, ½ tsp salt
¼ tsp of each - garam masala, amchoor and red chilli powder
2 tbsp yogurt/curd
8-10 kishmish

1. Grind onions, oil, dhania powder, haldi, garam masala and red chilli powder in a mixer to a smooth paste.
2. For the gravy, put onion paste in a deep microproof dish. Microwave for 8 minutes.
3. Add pureed tomatoes. Microwave for 7 minutes.
4. Add 1½ cups water. Microwave for 6 minutes. Keep aside.
5. For the koftas, mix carrots with all ingredients of the koftas except yogurt and kishmish.
6. Add yogurt. Mix well. Make 8 round balls with 1 kishmish stuffed in each. Place balls on a greased microproof plate in a ring and microwave for 3 minutes.
7. At serving time, place koftas in a serving dish. Pour curry over them. Microwave for 2 minutes and serve.

Makai-Mirch Salan

Baby corn and green chillies in a red gravy flavoured with cumin and mustard seeds.

Serves 4-5

5-6 big acchari hari mirch, 1 tbsp vinegar
200 gm babycorns - keep whole if small or cut into 2 pieces if big
3 tbsp oil
1 tsp jeera (cumin seeds)
½ tsp mustard seeds (rai)
a few curry leaves, 2 onions - chopped finely
2 tsp coriander (dhania) powder, 1 tsp salt, ¼ tsp red chilli powder
½ tsp dry mango powder (amchoor)
¼ tsp garam masala
1½ cups readymade tomato puree
1 tsp ginger paste
3 tbsp roasted peanuts
1 cup water, ¾ cup milk

Vegetable au Gratin : Recipe on Page 94 ➢

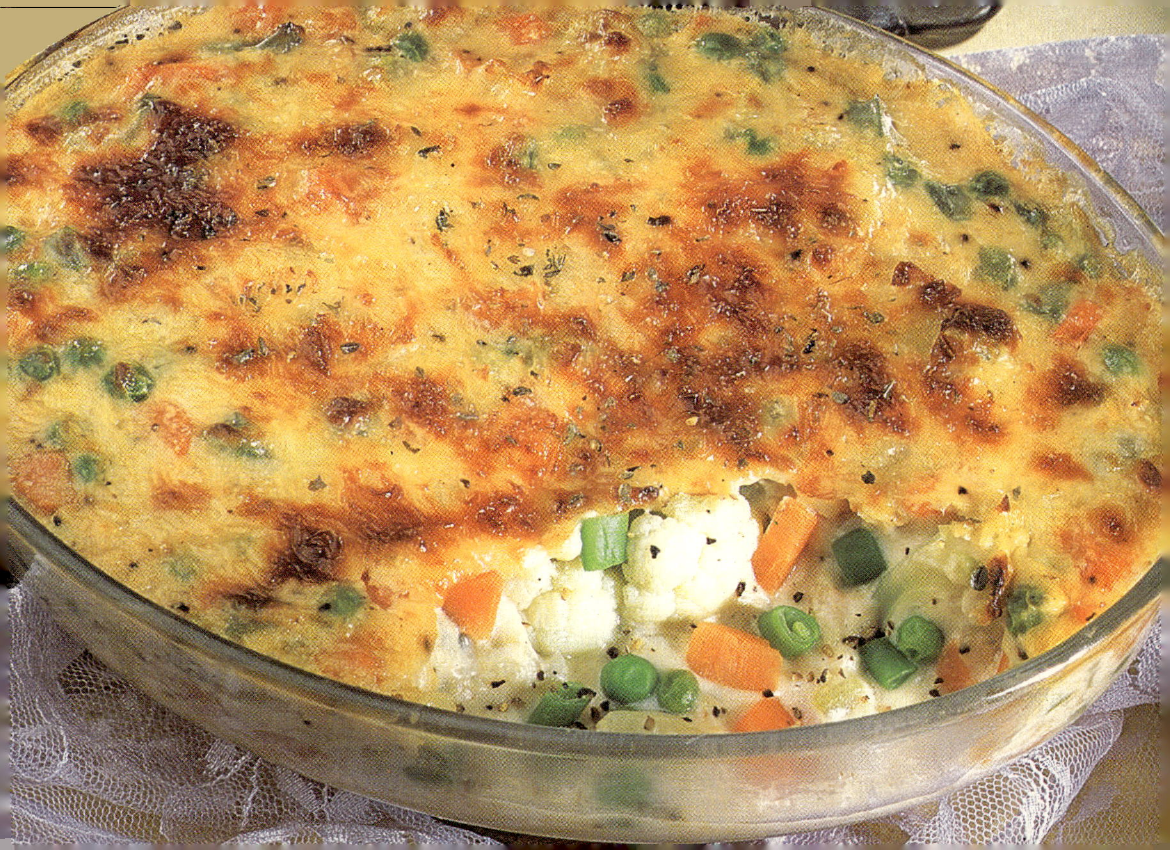

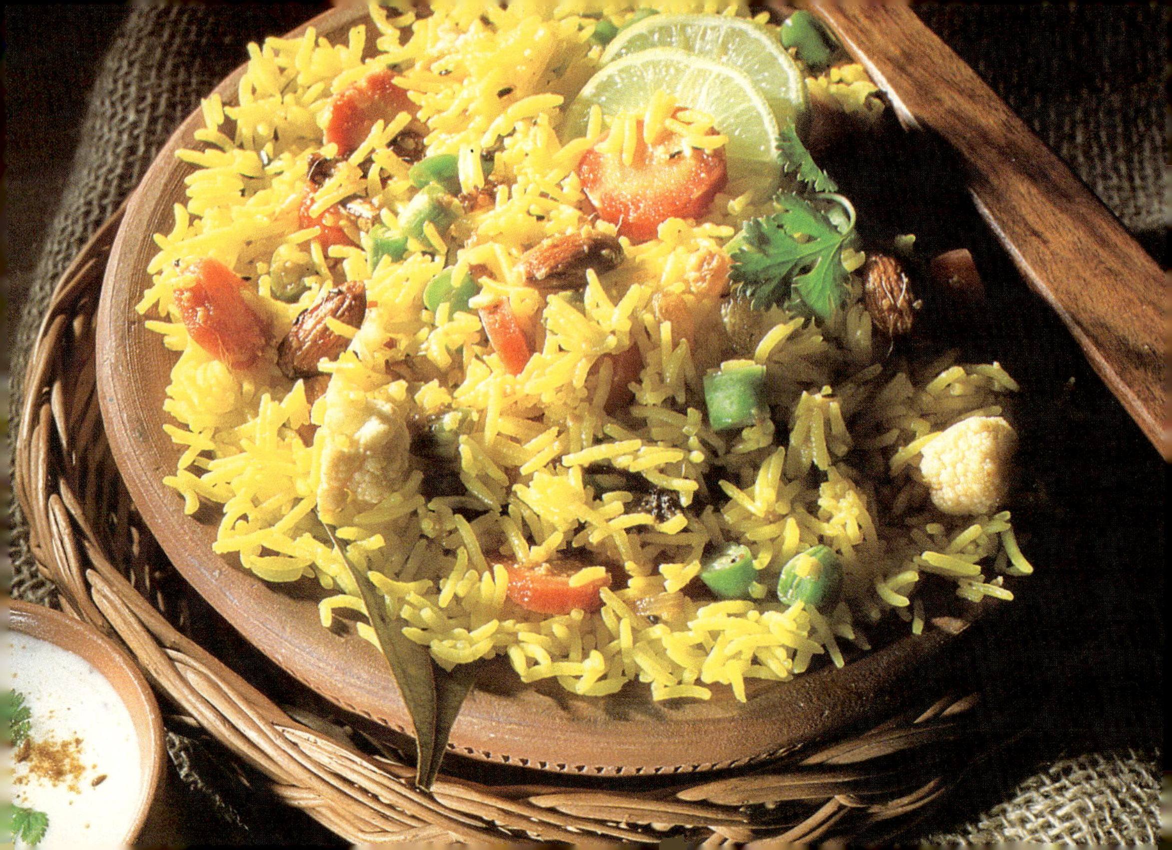

1. Slit the mirch and remove seeds. Sprinkle ½ tsp salt and 1 tbsp vinegar. Rub well and keep aside for 15 minutes. Wash and pat dry on a kitchen towel.
2. Churn peanuts with ¼ cup milk in a mixer to get a paste. Keep aside.
3. In a microproof dish put oil, jeera, rai, curry leaves, chopped onions, dhania powder, salt, red chilli powder, amchoor and garam masala. Mix well. Microwave for 8 minutes.
4. Add mirchi, baby corns, redymade tomato puree and ginger paste. Microwave for 6 minutes.
5. Add prepared peanut paste and 1 cup water. Mix well. Microwave for 5 minutes. Stir well.
6. Add ½ cup milk. Microwave for 1 minute. Serve hot.

◄ *Subz Pulao : Recipe on Page 80*

Paneer Pista Haryali

You can add anything else also, instead of paneer in this rich green gravy.

Serves 4

200 gm paneer - cut into 1" squares
2 medium sized onions - cut into 4 pieces
¼ cup pistas (pistachio nuts) with the hard cover on - remove hard cover
½ cup milk

GRIND TOGETHER TO A PASTE
1 green chilli - roughly chopped
¼ cup chopped fresh coriander
1" ginger piece and 4-5 flakes garlic
1 tbsp dhania powder (ground coriander)
½ tsp white pepper powder
¾ tsp salt, or to taste
4 tbsp oil

1. Peel and cut each onion into 4 pieces. Put onion pieces and pistas in 1 cup water in a microproof dish and microwave covered for 6 minutes. Cool slightly. Slip the skin of pistas.
2. Grind boiled onion pieces and the pistas along with the water, and with all the other ingredients written under paste to a fine green paste.
3. Put the prepared paste in a microproof dish and microwave for 5 minutes.
4. Add ½ cup water, a small pinch of sugar and paneer and microwave for 2 minutes. Keep aside till serving time.
5. At serving time, add ½ cup milk or slightly more to get a thick gravy. Microwave for 2 minutes. Serve hot.

Mixed Veggie Curry

Seasonal vegetables in a red tomato based gravy flavoured with cloves and cardamoms.

Serves 4

¼ of a small cauliflower - cut into 8 small ½" florets
1 carrot - cut into thin round slices
10 french beans - cut into ½" pieces
1 capsicum - cut into ½" cubes
50 gm paneer - cut into ½" cubes
¾ cup ready made tomato puree
1½ tsp salt, or to taste
1½ cups milk (cold)

GRIND TO A PASTE
2 onions, 2 tbsp ghee or oil
½ " piece ginger, 3-4 flakes garlic
2 laung (cloves)
seeds of 1 chhoti illaichi (green cardamom)

1 tsp dhania powder
¾ tsp jeera (cumin seeds) - crushed to a powder
½ tsp garam masala powder
¼-½ tsp red chilli powder

1. Cut all the vegetables into ½" pieces. Wash cauliflower, carrots and beans. Microwave together for 3 minutes in a plastic bag or a covered dish. Keep aside.
2. Grind together all ingredients of the paste. Put onion paste in the dish. Microwave uncovered for 8 minutes.
3. Add tomato puree, salt, all microwaved vegetables and capsicum. Mix well.
4. Microwave for 4 minutes.
5. Add paneer and milk. Mix well. Keep aside till serving time.
6. To serve, microwave for 3 minutes.

Khumb Matar Miloni

Mushroom and peas in a tomato - yogurt gravy.

Serves 4

1 packet (200 gm) mushrooms (khumb)
1 cup peas (shelled)
2 tbsp oil, 1 tsp ginger-garlic paste
1 tbsp fenugreek leaves (kasoori methi)

PASTE - 1

2 onions, 2 laung (cloves)
2 green cardamoms (chhoti illaichi)
1 tsp saunf (fennel)
¼ tsp turmeric powder, 3 tbsp oil

PASTE - 2

3 tomatoes - cut into 4 pieces
½ cup yogurt (dahi)
1¼ tsp salt, ½ tsp garam masala
½ tsp degi mirch or red chilli powder

1. Cut each mushroom into 4 pieces.
2. Put 2 tbsp oil, ginger- garlic paste and mushrooms in a microproof dish. Mix and spread them in the dish. Microwave for 3 minutes. Remove mushrooms from the dish and keep aside.
3. Grind all the ingredients of paste-1 in a mixer to a smooth paste.
4. Grind all the ingredients of paste-2 in a mixer to a smooth paste.
5. For the masala, put the paste-1 of onions in the same microproof dish and microwave for 7 minutes.
6. Add paste-2 of tomatoes and kasoori methi. Mix and microwave for 7 minutes.
7. Add 2 cups water and peas. Microwave for 6 minutes.
8. Add the mushrooms. Microwave for 2 minutes. Serve hot.

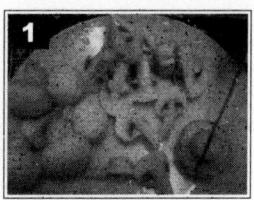

Special Sambar

The pulse is blended in a mixer to get a smooth and creamy sambhar.

Serves 4

½ cup arhar dal (red gram dal)
100 gm pumpkin or 2 small brinjals or any other vegetable of your choice - chopped (1 cup)
lemon sized ball of imli (tamarind)
1½ tsp salt or to taste
¼ tsp hing powder (asafoetida)
2 tbsp sambhar powder
1 tbsp oil
1 onion - sliced
½ tsp sarson (mustard seeds)
¼ cup curry leaves
tiny piece of gur (jaggery) - optional

1. Put dal in a microproof bowl. Add 1 cup water and microwave covered for 5 minutes. Remove the cover and microwave for 5 more minutes or till dal turns soft. Cool. Add ½ cup water. Mix. Blend in a mixer to a puree.
2. Microwave imli in ½ cup water for 2 minutes. Extract the juice. Add 1 more cup water to the left over imli and mash well. Extract more juice. Keep imli juice aside.
3. Put oil, curry leaves, sarson, hing, sambhar powder and onions in a deep microproof bowl. Mix well. Microwave for 5 minutes.
4. Add the chopped vegetables, salt, pureed dal and imli paani. Cover and microwave for 6 minutes.
5. Add 2 cups water and microwave for 8 minutes. Serve hot.

Paneer Makhani

Paneer in a red cashew based makhani gravy flavoured with fenugreek.

Picture on cover   Serves 4-5

300 gm paneer - cut into cubes
5 large (500 gm) tomatoes - chopped roughly
1" piece ginger - chopped
2 tbsp butter/ghee and 2 tbsp oil
seeds of 2 green illaichi (cardamoms) - crushed
½ tsp sugar, 1 tsp salt or to taste
½ tsp garam masala
½ tsp degi mirch or red chilli powder
1 tsp tomato ketchup
4 tbsp cashewnuts or magaz - soaked in ¼ cup water and ground to a paste
2 tsp kasoori methi (dried fenugreek leaves)
1 cup milk, approx.
3-4 tbsp cream

1. Microwave tomatoes and ginger in a deep dish with ½ cup water for 5 minutes.
2. Blend tomatoes and ginger to a puree in a mixer.
3. Microwave butter/ghee and oil for 2 minutes. Add illaichi powder. Mix. Add salt, sugar, red chilli powder and garam masala. Mix. Add fresh tomato puree and tomato ketchup. Mix very well. Microwave for 8 minutes. Stir once in between.
4. Add cashewnut or magaz paste and kasoori methi. Mix well. Add ½ cup water. Microwave for 3 minutes.
5. Add paneer and mix well. Add enough milk to get a thick red gravy. Mix well and microwave for 3 minutes.
6. Add cream. Sprinkle little kasoori methi on top and serve hot.

Chawal Bhare Tamatar

Tomato stuffed with crunchy rice and put in a gravy.

Serves 4

6 small firm tomatoes, 2 tbsp oil
2 tbsp tomato ketchup

FILLING

1½ cups cooked rice
1 tbsp roasted peanuts - roughly crushed
¼ cup chopped coriander, 2 tbsp grated cheese
2 green chillies - deseeded and chopped
1 tsp chaat masala, salt to taste, ½ tsp garam masala

GRAVY

2 big onions, 1" piece ginger
½ tsp (turmeric powder) haldi
1 tsp dhania (coriander) powder
½ tsp chilli powder, ¾ tsp salt
½ tsp garam masala, ¼ tsp amchoor

1. Slice a small piece from the top of each tomato. Scoop out carefully.
2. Rub some salt inside the tomatoes and keep them upside down.
3. Mix all ingredients of the filling. Do not mash. Mix gently.
4. Fill scooped tomatoes with the filling. Press well.
5. Grind all ingredients of gravy along with the scooped out portion of the tomatoes, together in a mixer.
6. Put oil and onion-tomato paste in a microproof bowl. Microwave for 11 minutes or more, till paste turns dry.
7. Add 1½ cups water and tomato ketchup. Mix well. Microwave for 6 minutes.
8. Arrange stuffed tomatoes on the gravy. Keep aside till serving time.
9. To serve, microwave for 3 minutes or till tomatoes turn soft.

Indian – Dry & Masala

Dal Maharani

Sukhi urad dal with each grain of dal standing out to perfection!

Serves 4

1 cup dhuli urad dal (split black beans) - soaked for 1 hour
1 onion - sliced, 1" piece ginger - grated, 3 tbsp oil
1¼ tsp salt, ½ tsp turmeric powder (haldi), ½ tsp red chilli powder
¼ tsp amchoor, ¼ tsp coriander (dhania) powder

1. Clean and wash dal. Soak in water for 1 hour.
2. Keep onion and ginger in a microproof dish. Sprinkle oil on it. Mix. Add salt, haldi, chilli powder, amchoor and dhania powder. Microwave for 6 minutes.
3. Drain the dal and add dal to the onions. Add 2 cups water. Mix well. Microwave covered for 20 minutes. Stir once after 8 minutes in-between.
4. After it is ready, let it stand for 3-4 minutes till it turns soft. Sprinkle chopped coriander and mix gently with a fork.

Grilled Besani Subzi

Serves 4 *Picture on page 1*

2 carrots - cut into thin round slices
2 capsicums - sliced into thin fingers
75 gm paneer - cut into thin fingers (¾ cup)
½ tsp ajwain (carom seeds), 3 tbsp besan (gramflour), 1 tsp lemon juice
¼ tsp red chilli powder, ¼ tsp dhania powder, ¼ tsp haldi
2 tsp channa masala, 2 tsp amchoor, 1 tbsp milk, 1 tsp salt
1 tomato - deseeded and cut into thin fingers

1. Microwave sliced carrots with ¼ cup water in a microproof dish for 3 minutes.
2. In another microproof dish put 3 tbsp oil, ajwain, besan, lemon juice, red chilli powder, dhania powder, haldi, channa masala and amchoor. Mix and microwave for 2 minutes.
3. Add carrot, capsicum, paneer, milk and salt.
4. Grill in the oven for 16 minutes. After 8 minutes, add deseeded tomatoes and mix gently with a fork and grill for the remaining 8 minutes. Serve hot.

Mili-Juli-Vegetable

Mixed vegetables flavoured with cardamoms.

Picture on facing page   Serves 4

1 big potato
200 gm (1 packet of 15- 20 pieces) baby cabbage (brussel sprouts) - trim the stalk end or use ½ of a small cabbage - cut into 1" pieces
100 gms baby corns (7-8) - keep whole
½ cup peas (matar)
1 carrot - cut into ¼" pieces (½ cup)
8-10 french beans - cut into ½" pieces
5-6 cherry tomatoes or 1 regular tomato - cut into 4, remove pulp

ONION PASTE (GRIND TOGETHER)
1 onion
2 cloves (laung)
seeds of 2 green cardamoms (illaichi)

Contd...

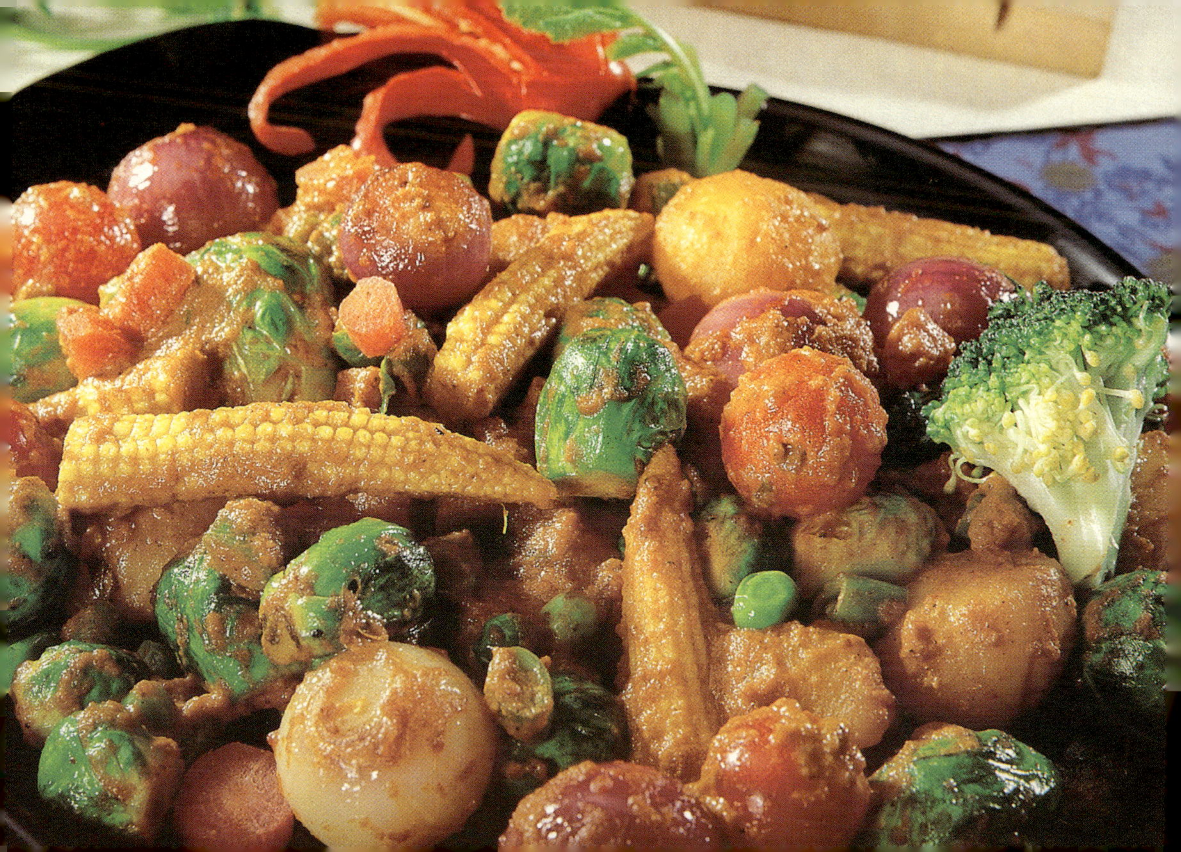

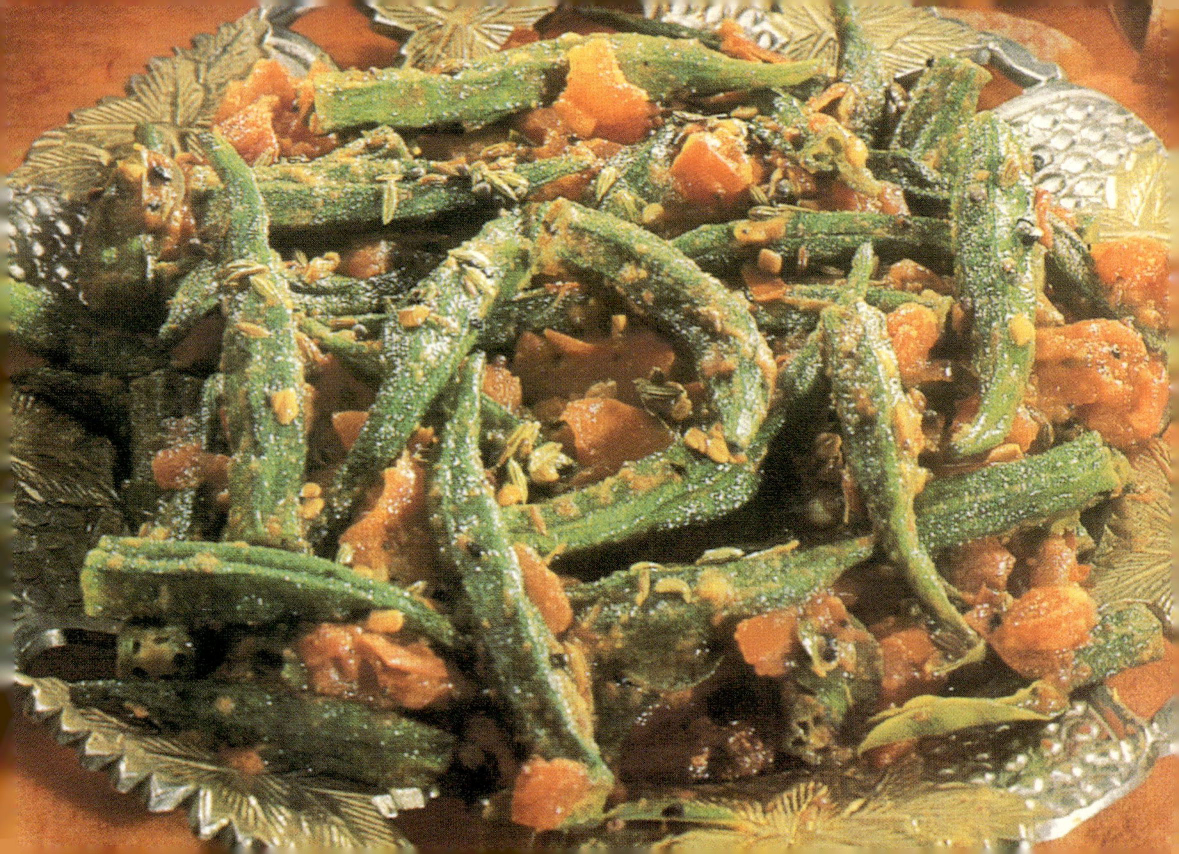

TOMATO PASTE (GRIND TOGETHER)
2 tomatoes
¼ cup curd
¼ tsp haldi, 1 tsp salt, ½ tsp chilli powder
½ tsp garam masala, ½ tsp degi mirch

1. Peel potato and make balls with the help of a melon scooper or cut eact potato into 8-10 small pieces with a knife.
2. Put 1 cup of water, 2 tsp salt, potato balls in a deep bowl and microwave for 5 minutes.
3. To the same water add cabbage, baby corns, peas, carrots and french beans. Microwave covered for 2 minutes. Strain.
4. Put 3 tbsp oil and onion paste in a microproof bowl. Microwave for 5 minutes.
5. Add tomato paste. Mix. Microwave for 7 minutes.
6. Add ½ cup water and vegetables. Mix well. Microwave covered for 3 minutes. Serve hot.

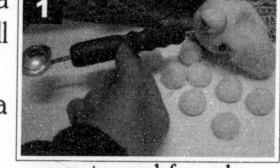

◄ *Achaari Bhindi : Recipe on Page 68*

Achaari Bhindi

Crispy fried lady's fingers with pickle spices.

Picture on page 67 Serves 4

500 gm bhindi (lady's finger)
1 tsp ginger paste
½ tsp red chilli powder
1 tsp dhania powder
½ tsp amchoor, ½ tsp garam masala
¾ tsp salt, or to taste
2 big tomatoes - chopped
1 tsp lemon juice

ACHAARI SPICES
a pinch of hing (asafoetida)
1 tsp saunf (fennel), ½ tsp rai (mustard seeds)
½ tsp kalonji (onion seeds)

1. Wash bhindi and wipe dry. Cut the tip of the head of each bhindi, leaving the pointed end as it is. Now cut the bhindi lengthwise making 2 thin long pieces from each bhindi.
2. Keep bhindi in a dish. Sprinkle 2 tbsp oil on it. Mix well. Put them in the oven on combination mode (micro+grill) for 15 minutes or till cooked and crisp. Keep aside.
3. In the separate small dish put 2 tbsp oil and achari spices. Microwave for 3 minutes.
4. To the bhindi, add achari spices, red chilli powder. dhania powder, amchoor, garam masala, dry masala powders, salt, ginger paste, tomatoes and lemon juice. Mix very well. Microwave for 4 minutes. Serve hot.

Crispy Achaari Mirch

Peppers filled with pickle masala rice and grilled with a semolina coating till crisp.

Picture on page 1 *Serves 6*

125 gms (6) big, fat green chillies (achari hari mirch) or 3 small capsicums

FILLING

1½ cups boiled rice, ¼ cup vinegar
½ tsp brown mustard seeds (rai)
½ tsp jeera (cumin seeds), ½ tsp saunf (fennel seeds)
1 onion - chopped
¼ tsp haldi (turmeric powder)
2 tsp of any achaar ka masala (preferably use aam ka achaar)
1 tomato - chopped
½ tsp salt, 1 tsp tomato ketchup

COATING

2 tbsp maida (flour), 4 tbsp suji (semolina)
¼ tsp salt
¼ tsp garlic paste

1. Slit the chillies and remove the seeds. Pour vinegar on them and sprinkle ¼ tsp salt on them. Mix. Keep aside.
   If using capsicum cut a thin slice from the top of the peppers. Pull out the stalk end and make it hollow. Rub a little salt, lemon juice and ginger paste on the inner surface. Keep them upside down. Pour some oil on the outer surface. Rub the oil on the outer and inner surface and keep aside for 10 minutes.
2. In a microproof dish put 1 tbsp oil, jeera, rai, saunf, chopped onion, and haldi. Mix well. Microwave for 5 minutes.
3. Add boiled rice, aam ke aachar ka masala, ½ tsp salt, chopped tomato and tomato ketchup. Mix well.
4. Fill each mirchi with the filling. Fill as much as the mirchi can take.
5. Mix ingredients of coating in a plate.
6. Put 2-3 tbsp oil in a bowl. Dip the sides of the mirchi in the oil and then immediately roll over the coating spread in the plate. Coat all the sides of the mirchi with the coating mixture nicely.
7. Grill for 12-15 minutes or till golden. Serve hot.

Zayekedar Arbi

Serves 3-4

10 medium sized pieces of arbi (400 gm) - peeled & halved lengthways
1" piece ginger - grated finely
4-5 flakes garlic - chopped & crushed
1-2 green chillies - chopped very finely
¾ tsp salt
juice of ½ lemon
1 tsp ajwain - crushed roughly

MASALA
6 tbsp thick curd
1 tsp besan (gram flour)
1 tbsp fresh coriander chopped
1 tbsp (poodina) mint leaves - chopped
2 tsp oil
½ tsp salt, ½ tsp red chilli powder, ½ tsp garam masala, ½ tsp amchoor

1. Choose even sized arbi & it should not be too thick. Peel, slit into half length ways. If the arbi is thick, cut lengthways into 3 pieces. Rub 2 tsp salt on it nicely and keep aside for 10 minutes or more. Wash well. Strain. Wipe dry.
2. Place arbi in a flat dish. Sprinkle all the ingredients given under the arbi on it. Rub them well over the arbi.
3. Microwave 2 minutes uncovered. Turn pieces over with tongs or spoon. Again microwave 2 minutes uncovered.
4. Beat curd with a spoon in a small bowl. Add all the other ingredients of the masala and mix well.
5. Sprinkle the dahi mix on the arbi and mix well. Microwave covered for 6 minutes.
6. Let it stand for 2 minutes. Serve sprinkled with a little lemon juice.

Baingan ka Bharta

Picture on facing page   *Serves 3-4*

1 medium baingan (brinjal) of round variety (350 gm)
2 onions - chopped finely
½ cup ready-made tomato puree, 1 tomato - chopped
½" piece ginger - chopped finely, 1 green chilli - chopped
2 tsp dhania (coriander) powder, ½ tsp garam masala
½ tsp degi mirch or red chilli powder, 1 tsp salt, ¼ tsp haldi

1. Place brinjal in a microproof flat dish. Microwave for 5 minutes. Let it cool down.
2. Cut brinjal into half and scoop out the pulp. Mash the pulp with a fork and keep pulp aside.
3. In the same dish, put 3 tbsp oil, onions, ginger, green chilli, dhania powder, haldi, garam masala, degi mirch & microwave for 7 minutes.
4. Add brinjal pulp and cook on combination mode (micro+grill) for 10 minutes.
5. Add chopped tomato and tomato puree and 1 tsp salt. Mix well. Microwave for 6 minutes. Serve hot.

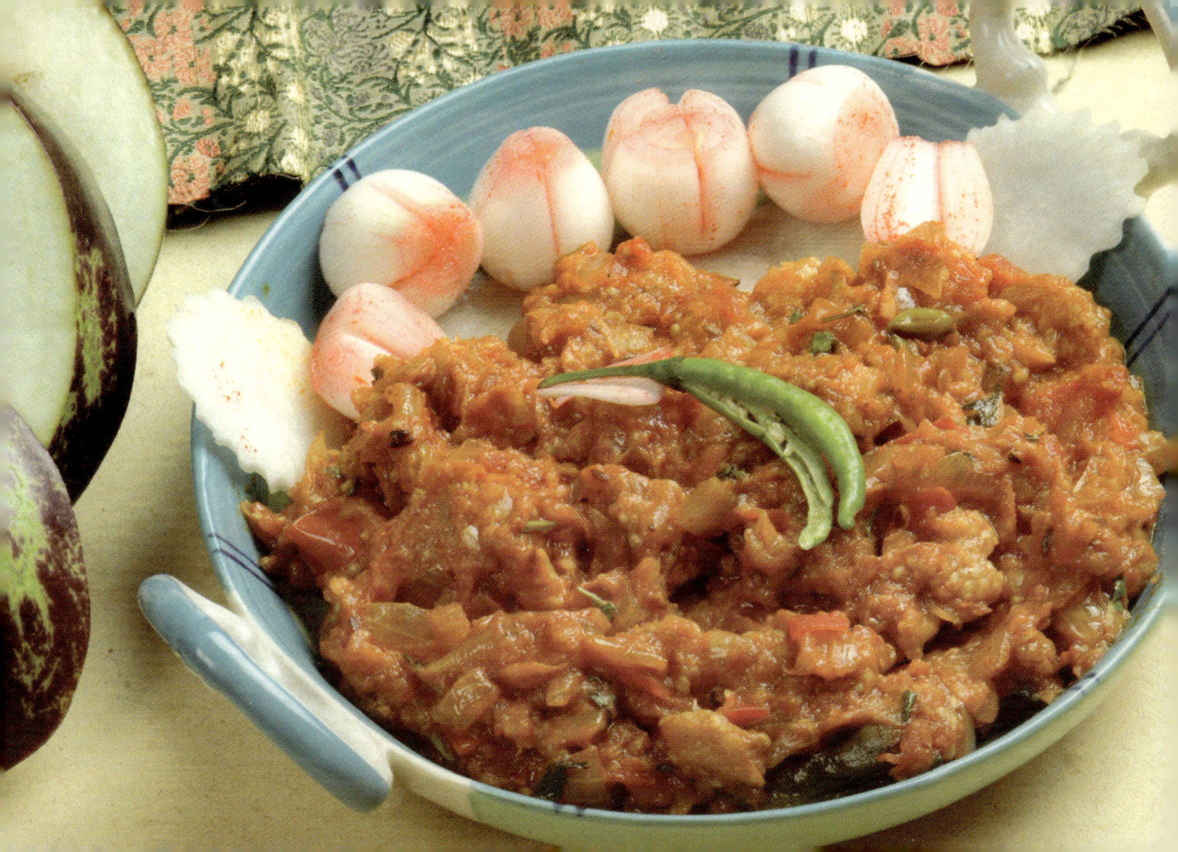

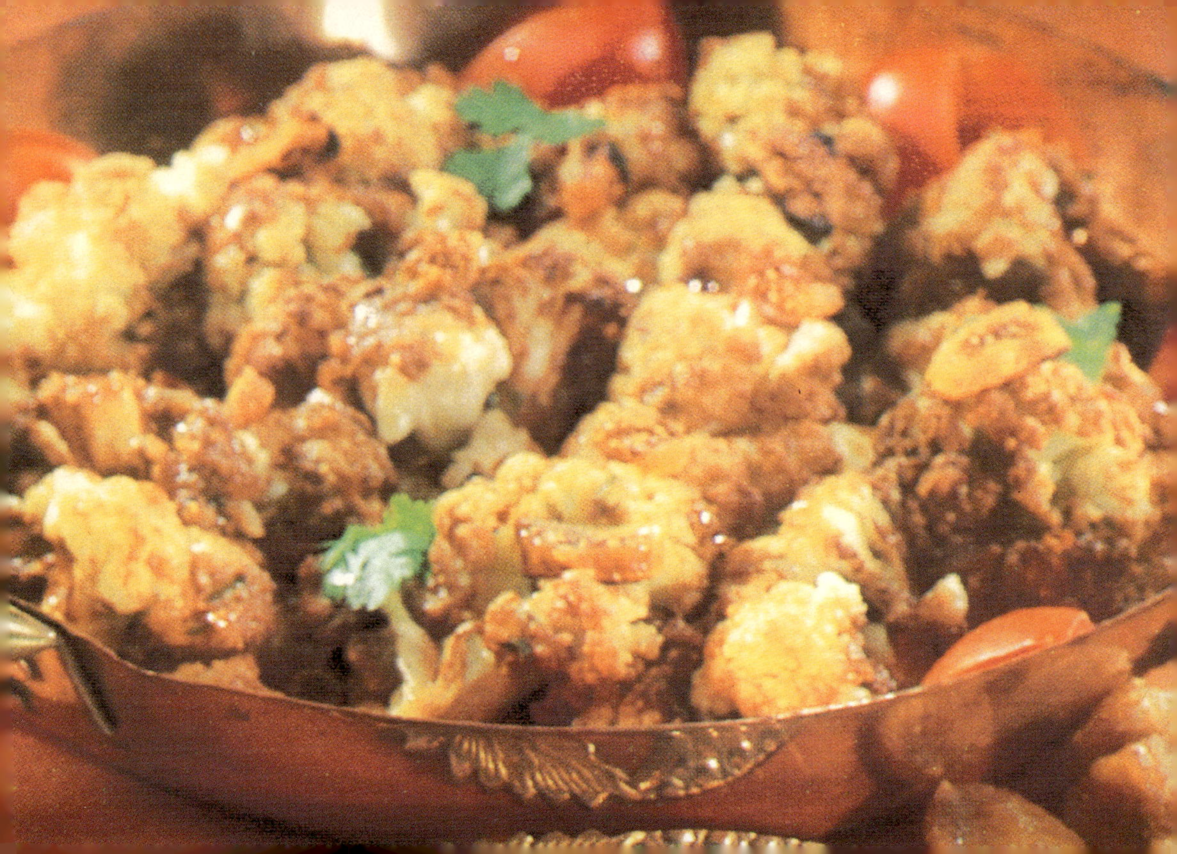

Anjeeri Gobhi

Cauliflower cooked with a hint of sweetness in a yogurt and dry figs paste.

Serves 4-6   Picture on opposite page

(1 big) ½ kg cauliflower (gobhi) - cut into medium size florets with long stalks
1 tsp jeera (cumin seeds), 2 onions - chopped, ¾" piece ginger- chopped
¼ tsp turmeric (haldi), 2 green chillies, 1 tomato - chopped

ANJEER PASTE

8 small anjeers (figs) - chopped, ¾ cup dahi (yogurt), ½ tsp garam masala
½ tsp red chilli powder, 1½ tsp salt

1. Break the cauliflower into medium florets, keeping the stalk intact.
2. Churn all the ingredients given under anjeer paste in a mixer till smooth.
3. In a microproof dish put 4 tbsp oil, jeera, chopped onions and ginger. Add haldi. Mix. Microwave for 9 minutes.
4. Add the prepared anjeer paste. Mix well. Add cauliflower and mix very well. Mix in whole green chillies and chopped tomato. Cover and microwave for 10 minutes or more till the cauliflower gets cooked.

Kadhai Paneer

Paneer with capsicum strips on in tomato masala flavoured with fenugreek.

Serves 4

200 gms paneer - cut into thin fingers
3 tbsp oil, 5-6 flakes garlic - crushed
½ cup ready made tomato puree
1 tsp kasoori methi (dry fenugreek leaves)
1 tsp salt, ½ tsp sugar
¾ tsp red chilli powder (to taste), 1 tsp dhania powder, ½ tsp garam masala
1 capsicum - cut into thin long strips

1. In a dish add oil and garlic. Microwave uncovered 2 minutes.
2. Add tomato puree and kasoori methi. Add salt, sugar, red chilli powderd, dhania and garam masala. Mix well. Add capsicum. Mix. Microwave uncovered for 3 minutes.
3. Add paneer. Mix well. Keep aside. At serving time, microwave covered for 2 minutes. Serve hot in a small copper kadhai.

Rice

Steamed Rice

Rice prepared in the microwave should not be served immediately as it tastes uncooked if done so. Microwaved rice is wonderful, each grain is soft and separate, if served after 10 minutes, once the microwave is shut off.

Serves 2

1 cup basmati rice - washed and soaked for ½ hour
½ tsp salt
1 tsp lemon juice
2 cups water

1. Drain the soaked rice. Put in a broad, shallow microproof dish (a pie dish).
2. Add salt, lemon juice and 2 cups water. Mix well.
3. Microwave covered for 11 minutes. Serve after 10 minutes.

Subz Pullao

The spice bag added to rice and the vegetables, makes it very aromatic.

Picture on page 48 Serves 3-4

1 cup basmati rice - washed and soaked
5 tbsp oil
1 tsp ginger paste, ½ tsp garlic paste
¼ tsp haldi, ½ tsp red chilli powder
1½ tsp salt or to taste

SABOOT MASALA OR SPICE BAG (*crush together and tie in a piece of muslin cloth*)

10 saboot kali mirch (pepper corns)
2 tsp saunf (fennel seeds)
3-4 chhoti illaichi (green cardamom)
3-4 moti illaichi (black cardamom)
4 laung (cloves)
2 sticks dalchini (cinnamon)

VEGETABLES

1 potato - cut into ½" pieces
½ of a small cauliflower - cut into small florets
2 carrots - cut into ½" pieces
1 cup green peas (matar)
1 tomato - cut into 8 pieces
2-3 green chillies - cut into thin strips
1 tbsp mint leaves (poodina), 1 tsp lemon juice

1. In a broad microproof dish, put oil, ginger paste, garlic paste, cauliflower, potato, carrots and peas. Microwave for 3 minutes.
2. Drain the soaked rice. Add rice, 2 cups water, haldi, red chilli powder and salt. Add the spice bag. Microwave covered for 6 minutes.
3. Add tomatoes, mint, coriander, green chillies and lemon juice. Stir gently with a fork. Cover and microwave for 7 minutes. Wait for 5 minutes. Fluff with a fork. Remove spice bag. Serve.

Chinese And Thai

Honey Chilli Veggies

Sweet and spicy mixed vegetables with Chinese sauces.

Picture on page 30 *Serves 4*

1 large carrot
8-10 mushrooms - keep whole
8-9 baby corns - keep whole if small and divide into two lengthwise, if thick
1½ cups cauliflower or broccoli - cut into small, flat florets
1 onion - cut into 8 pieces
1 capsicum - cut into ½" cubes
4 tbsp oil
2-3 dry, red chillies - broken into bits and deseeded
15 flakes garlic - crushed
¾ tsp salt and ¼ tsp pepper, or to taste

a pinch ajinomoto (optional)
1½ tbsp vinegar, 1 tsp soya sauce
2½ tbsp tomato ketchup
2-3 tsp red chilli sauce
3-4 tsp honey, according to taste
3 tbsp cornflour dissolved in ½ cup water alongwith 1 seasoning cube

1. Dissolve cornflour in ½ cup water. Add seasoning cube and keep aside.
2. Put oil, broken red chillies, garlic, baby corns, mushrooms, carrots, cauliflower and onion in a microproof dish. Mix well. Microwave for 5 minutes.
3. Add pepper, salt, ajinomoto, chilli sauce, tomato sauce, soya sauce, honey, and vinegar. Mix and microwave for 1 minute.
4. Add capsicum and dissolved cornflour and mix. Microwave for 3 minutes or till the vegetables are crisp-tender and the sauce coats the veggies. Mix well before serving. Serve hot with rice or noodles.

Broccoli in Butter Sauce

Broccoli in the new white Chinese sauce prepared from butter and milk and thickened with flour.

Picture on facing page   *Serves 4*

250 gm (1 medium head) broccoli
1 tsp salt, 1 tsp sugar

SAUCE

1 veg seasoning cube (maggi or knorr)
3 tbsp butter
1 onion - sliced
1 tbsp crushed garlic (15 flakes)
1 tbsp chopped coriander
3 tbsp flour (maida)
1 cup milk
2 tsp mustard paste
½ tsp pepper, ¾ tsp salt, or to taste
1 cup thin cream

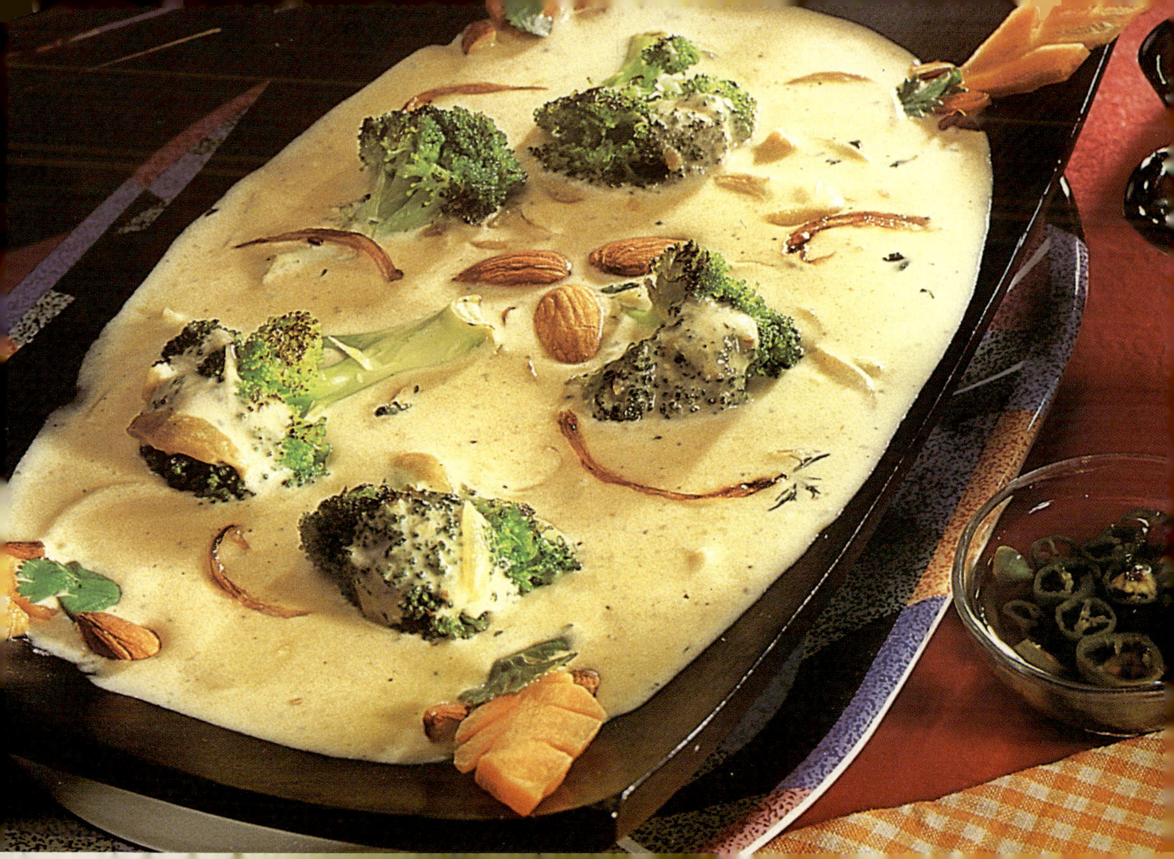

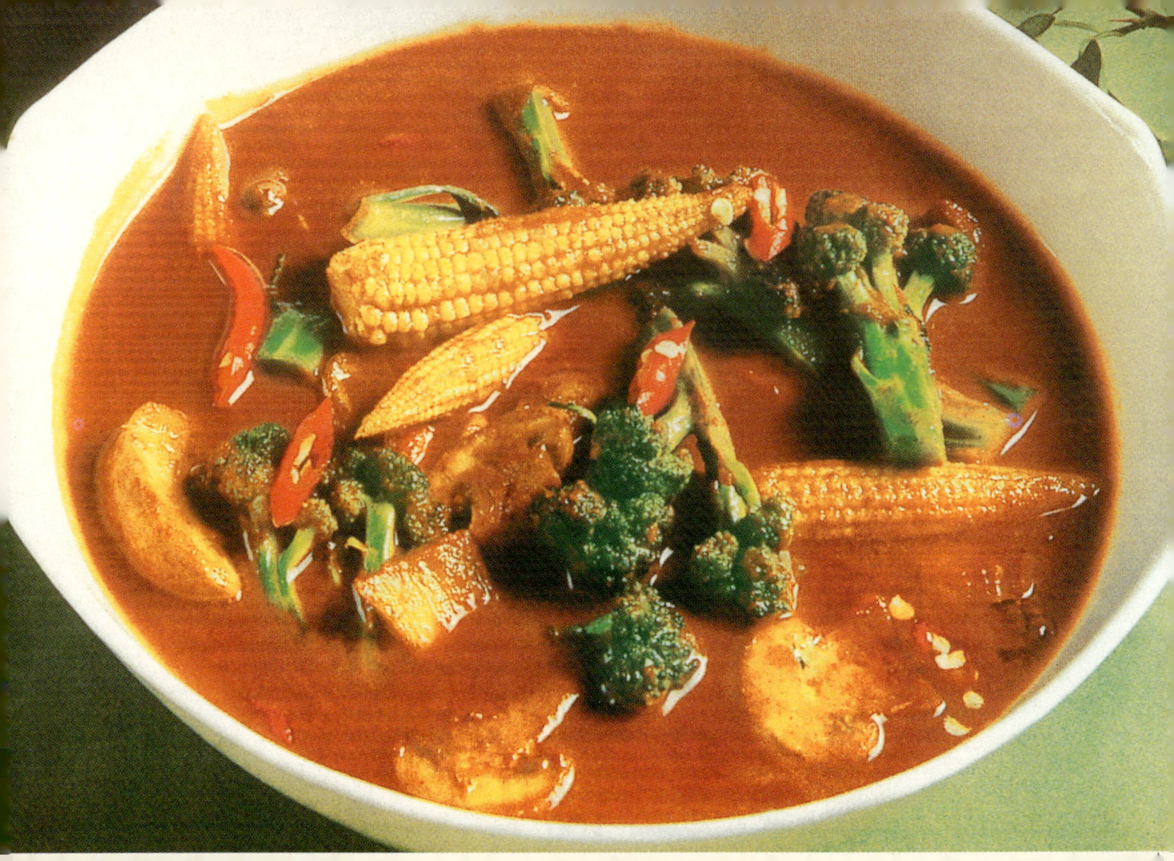

1. Cut broccoli into medium sized florets with long stalks.
2. Put 1 cup water in a microproof bowl. Add 1 tsp salt and 1 tsp sugar and mix. Add broccoli to it and mix well. Microwave covered for 3 minutes. Drain. Refresh in cold water. Wipe dry broccoli with a clean kitchen towel.
3. Put 3 tbsp butter in a microproof dish and microwave for 30 seconds. Add sliced onion, crushed garlic and microwave for 5 minutes.
4. Add broccoli, coriander, crushed seasoning cube and maida. Mix and microwave for 1 minute.
5. Add milk, ¾ cup water, mustard paste, pepper and salt. Mix and microwave for 6 minutes or till sauce thickens. Stir once in between. Remove.
6. Add cream. Mix. Keep aside till serving time. At serving time, microwave for 2 minutes.

◄ *Veggie Red Thai Curry : Recipe on Page 88*

Veggie Thai Red Curry

Mixed vegetables in a lemon flavoured, spicy red curry prepared from coconut milk.

Picture on page 86   *Serves 4-6*

RED CURRY PASTE

4-5 dry, Kashmiri red chillies - soaked in ½ cup warm water for 10 minutes
½ onion - chopped, 8-10 flakes garlic - peeled
1½" piece ginger - chopped
1 stalk lemon grass or rind of 1 lemon
1½ tsp coriander seeds (dhania saboot)
1 tsp cumin seeds (jeera)
6 peppercorns (saboot kali mirch)
1 tsp salt, 1 tbsp vinegar, 2 tbsp oil

VEGETABLES

7-8 baby corns - slit lengthwise
2 small brinjals - peeled and diced or 8 French beans - cut into 1" pieces
1 small broccoli or ½ cauliflower - cut into small florets
7-8 mushrooms - sliced

OTHER INGREDIENTS
2½ cups ready made coconut milk
½ tsp soya sauce
2 tbsp chopped basil (tulsi) or coriander
salt to taste, ½ tsp brown sugar or regular sugar

1. Grind all the ingredients of paste with the water in which the chillies were soaked, to a very fine red paste.
2. Put the prepared red paste in a microproof dish. Microwave for 3 minutes.
3. Add ½ cup of coconut milk, vegetables and microwave for 4 minutes.
4. Add the rest of the coconut milk, soya sauce and chopped basil. Mix and microwave for 4 minutes.
5. Add salt and sugar to taste. Microwave for 1 minute. Serve hot with steamed rice.

Glass Noodles with Sesame Paste

Serves 6

100 gms glass noodles or rice seviyaan
2 tbsp oil
3 spring onions - cut into rings, till the greens, keep white separate

SESAME PASTE (GRIND ALL TOGETHER)

3 tbsp sesame seeds (til) - soak for 10 minutes in 5 tbsp hot milk & 2 tbsp water
½ tsp red chilli powder or to taste
¾ tsp salt
4 flakes garlic - finely chopped
1½ tbsp soya sauce
½ tsp sugar

1. Cut white spring onion into rings till the greens.
2. In a deep bowl microwave 3 cups water with 1 tsp salt and 1 tsp oil for 8 minutes. Add noodles to hot water. Cover and keep aside for 5 minutes in hot water. Drain noodles.
3. Wash with cold water several times. Strain. Leave them in the strainer for 15-20 minutes, turning them upside down, once after about 10 minutes to ensure complete drying. Apply 1 tsp oil on the noodles and spread on a large tray. Dry the noodles under a fan for 15-20 minutes. Keep aside till further use.
4. Grind all ingredients of sesame paste to a smooth paste.
5. Mix oil and white of spring onions in a microproof dish and microwave for 3 minutes.
6. Add prepared sesame mixture, mix well and microwave for 1 minute.
7. Add noodles, mix well. Add spring onion greens. Mix.
8. Microwave for 1 minute at serving time.

Baked Dishes

Macaroni Alfredo

Macaroni with vegetables cooked in cheese sauce. Tastes best if grilled till golden brown.

Serves 5-6

1 cup uncooked macaroni
100 gm mushrooms - sliced
50-100 gm baby corns - sliced (optional)
2 tbsp butter, 1 tsp oregano
1 onion or 2 spring onions - chopped along with the green parts
2½ tbsp flour (maida)
1¾ cups milk
¾ tsp salt, or to taste, ½ tsp pepper
½ tsp red chilli flakes
100 gm mozerrela cheese - grated
2 tbsp bread crumbs
some tomato slices and chopped parsley

1. Put 1½ cups of water and 1 tsp oil in a deep microproof bowl. Microwave uncovered for 3 minutes.
2. Add macaroni. Mix. Microwave uncovered for 5 minutes. Let it stand in hot water for 4-5 minutes. Drain and wash well with cold water.
3. In another microproof flat dish, microwave butter for 30 seconds.
4. Add oregano, spring onions, mushrooms and baby corns. Microwave uncovered for 5 minutes.
5. Add flour. Mix well and microwave uncovered for 30 seconds.
6. Add milk, salt, pepper and chilli flakes Mix and microwave uncovered for 6 minutes, stirring once in between. Microwave for 1-2 minutes more if the sauce does not turn thick.
7. Add macaroni and ½ of grated cheese. Mix well. Sprinkle bread crumbs. Arrange tomato, chopped parsley or coriander and the left over grated cheese.
8. At serving time, grill for 10-12 minutes or more till the top turns golden.

Vegetable au Gratin

Mixed vegetables baked in a cheese sauce topped with bread crumbs and tomato slices.

Picture on page 47   *Serves 8*

WHITE SAUCE

4 tbsp butter, 4 tbsp maida (plain flour)
2½ cups milk
1 tbsp tomato ketchup
salt, pepper to taste

VEGETABLES

10-15 french beans - cut into ¼" pieces
2 carrots - cut into small cubes
½ small cauliflower - cut into ½" florets
½ cup shelled peas
1 medium potato - cut into small cubes
½ of small ghiya (bottle gourd) - peeled and cut into small cubes (1 cup)

TOPPING
¼ cup bread crumbs
1 tomato - sliced

1. To prepare the sauce, melt butter for 50 seconds in a microproof dish.
2. Add flour, salt, pepper, tomato ketchup. Microwave for 30 seconds.
3. Add milk. Mix well. Microwave for 6 minutes. Keep sauce aside.
4. Wash vegetables and put in a microproof deep bowl with 1 tsp salt and ¼ cup water. Microwave covered for 5 minutes.
5. Mix vegetables with the prepared sauce. Add salt if required. Microwave for 3 minutes or till sauce turns thick and coats the vegetables.
6. Arrange tomato slices over it. Sprinkle bread crumbs.
7. Set microwave oven at 200°C using the oven (convection) mode and press start to preheat oven. Put the vegetables inside the hot oven and re-set the preheated oven again at 200°C for 20-25 minutes. Bake till golden brown. Serve hot.

Rice-Vegetable Ring

Saucy vegetables surrounded by a ring of rice mixed with green herbs.

Picture on page 19   Serves 5-6

RICE (MIX TOGETHER)
2 cups cooked rice
¼ cup chopped parsley or coriander
salt and lemon juice to taste

OTHER INGREDIENTS
2 tbsp butter
100 gm baby corns - sliced
2 cups finely chopped spinach
2½ tbsp flour (maida)
2 cups milk
¾ tsp salt, 1 tsp pepper, 1½ tsp oregano
½-1 cup grated mozzarella cheese (50-100 gm)
some tomato slices and black olives

1. Melt butter in a microproof dish for 40 seconds.
2. Add oregano, spinach and baby corns Microwave uncovered for 5 minutes.
3. Add flour. Mix well and microwave uncovered for 1 minute.
4. Add milk, salt, 2 tbsp cheese and pepper. Mix. Microwave uncovered for 5 minutes. Keep vegetables aside.
5. Spread parsley rice in a greased dish. Push rice toward the edges of the dish to get a rice border.
6. Sprinkle some cheese on it. Leaving aside the border of rice put the vegetables in the center portion of the dish, such that the rice border forms a ring around the vegetables.
7. Arrange tomato slices around the vegetables and sprinkle the left over grated cheese. Sprinkle olives.
8. To serve, microwave for 4 minutes or grill in the oven for 15 minutes.

Desserts And Cakes

Gajar ka Halwa

Carrot halwah made in very little fat and in a jiffy too.

Serves 5-6

½ kg carrots - grated
1½ cups milk
½ cup sugar - powdered, or to taste
½ cup (100 gms) khoya - grated
2-3 tbsp desi ghee
some chopped nuts like almonds, raisins (kishmish) etc.

1. Mix grated carrots and milk in a big deep bowl.
2. Microwave uncovered for 15 minutes. Mix once after 5 minutes.
3. Add sugar and khoya. Mix well.
4. Microwave for 10 minutes uncovered.
5. Add ghee. Mix well. Microwave for 7 minutes. Mix chopped nuts. Serve hot or cold decorated with nuts.

Eggless Cake with Mocha Icing

A quick microwaved chocolate cake topped with chocolate icing flavoured with coffee.

Picture on back cover *Serves 6*

½ tin condensed milk (milk-maid)
½ cup milk, ½ cup (75 gm) butter
1½ tbsp powdered sugar
100 gms (1 cup) maida (plain flour)
¼ cup cocoa, ¾ tsp level soda-bicarb
¾ tsp level baking powder
1 tsp vanilla essence

MOCHA GLAZE ICING

4 tbsp cocoa powder, 2 tbsp butter - softened
1 tsp coffee
1 cup icing sugar - sifted
2-3 tbsp chopped walnuts (akhrot)

TO SOAK

¼ cup coke or any other cola drink

1. Sift maida with cocoa, soda-bi-carb and baking powder. Keep aside.
2. Mix sugar and butter. Beat till very fluffy. Add milk-maid. Beat well.
3. Add milk and essence. Add maida. Beat well for 3-4 minutes till the mixture is smooth and light. Transfer to a big, greased deep dish of 8" diameter.
4. Microwave for 5 minutes. Let it cool.
5. Cut cake into two halves and soak each piece with cola drink on the cut side.
6. For the icing, microwave 4 tbsp water in a bowl for 1 minute. Add coffee and mix. Add cocoa and butter to it and mix well. Return to microwave and microwave at 70% power for 1 minute or till butter melts. Gradually add the sifted icing sugar. Mix well.
7. Glaze the cake with the icing, making peaks with the spoon. Decorate it with walnuts.
8. Refrigerate until glaze is set. Serve with ice cream.

Chocolate Walnut Cake (Eggless)

Picture on page 103   *Serves 6*

½ tin condensed milk (milkmaid)
½ cup milk
½ cup butter (75 gm)
1½ tbsp powdered sugar
100 gm (1 cup) maida (palin flour)
¼ cup cocoa
¾ tsp level soda-bi-carb (mitha soda)
¾ tsp level baking powder
1 tsp vanilla essence
2-3 tbsp chopped walnuts - mixed with 1 tsp maida

TOPPING
some ready-made chocolate sauce
2-3 walnuts - halves

Contd...

1. Sift maida, cocoa, soda-bi-carb and baking powder.
2. Mix sugar and butter. Beat with an electric hand mixer till fluffy.
3. Add milkmaid and beat some more, for about 2 minutes.
4. Add milk and essence. Add maida. Beat well for about 3-4 minutes, till the mixture is smooth and light.
5. Mix walnuts mixed with 1 tsp of maida and add to the cake batter.
6. Grease a deep, big bowl. Transfer mixture to the bowl. Microwave for 5 minutes. The cake will appear wet after 5 minutes, but just let it be undisturbed in the microwave for 10 minutes. (Do not microwave more)
7. Remove cake from bowl on the serving plate. Pour some chocolate sauce on the hot cake. Sprinkle some more walnuts if you like. Serve.

Vanilla Cake

For vanilla cake, add ¼ cup cornflour to maida and remove ¼ cup cocoa. The rest of the recipe is the same.

Chocolate Walnut Cake (Eggless): Recipe on page 101 ➤

BEST SELLERS BY SNAB
Excellence in Books

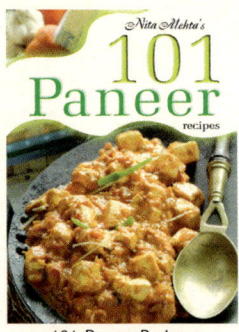

- 101 Paneer Recipes
- 101 Vegetarian Recipes 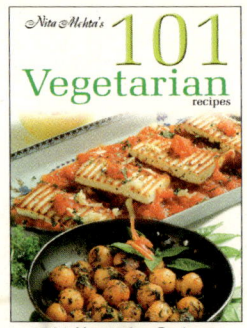
- SPECIAL Vegetarian Recipes 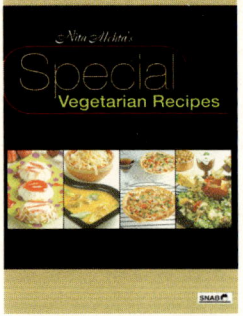
- Cakes & Cake Decorations 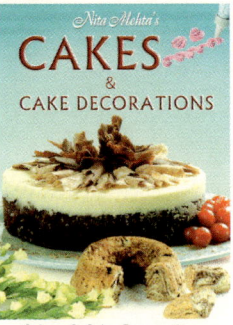
- DIABETES Cookbook 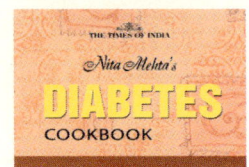
- Burgers & Sandwiches 
- Vegetarian MUGHLAI 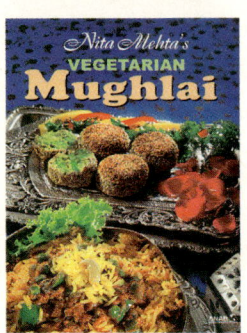
- CHOCOLATE Cookbook